WEATHER WARS

SHAUN MARSHALL

Dedicated to Robert, my amazing brother, for all the adventures we should have had, but never did.
I miss those journeys we never took, and I carry the thought of them with me always.

To Helen, for a shared vision of the universe, the unknown, the mysteries

The tour never ends.

ACKNOWLEDGEMENTS

To (Indi) Helen, my fellow traveller through mythology, conspiracies, and the vast unknown. Thank you for sharing with me a vision of the universe that goes beyond the ordinary. For the 9/11 talks, UFOs, the deep conversations, the unseen, the unexplained, the videos we shared, (heckel fish), and all the wonders in between. Your curiosity, your insight, and most of all, your company in exploring these hidden corners of life have made the journey brighter and infinitely more exciting. The truth is out there!

To Anthony, my star child. You're my constant rock, my guiding light, and the one who always encourages me, whether my ideals are good, bad, or somewhere in between. Your belief never wavers, and for that I am forever grateful.

TABLE OF CONTENTS

THE FORECAST

The news began without the usual music. No smiling, weather girl. No light headlines. Just the weather presenter, standing in front of a world map bathed in violent reds, blues, and warning symbols.

"This is the BBC weather forecast for Tuesday, 28th March 2029.
Morning. We are now entering the 36th day of extreme weather across the United Kingdom and much of the world. Here at home, temperatures are expected to exceed the high thirties and again today—the fourth time this week. The past two months have now been confirmed as the hottest ever recorded in British history, topping out at 42 degrees!"

Her tone was crisp, controlled, but her eyes told another story.

"Water rationing is now mandatory across the entire country. Multiple reservoirs have dropped below critical levels or are completely empty. Hospitals are operating at capacity with heatstroke, dehydration, and respiratory

emergencies. *More than two hundred schools in the south will remain closed indefinitely. Citizens are advised to avoid unnecessary travel and remain indoors during peak heat hours, from 11 a.m. to 5 p.m."*

The map shifted, showing fire icons scattered across the southern half of England.

"Wildfires are burning in Hampshire, Kent, and Dorset. Evacuations are underway in several villages. Across Europe, flames have consumed over a million hectares of forest, while thick smoke blankets entire regions, reducing visibility to a few metres in some areas."

She drew a breath, steadying herself before continuing,

"In the southern hemisphere, an entirely different crisis is unfolding. Temperatures in parts of Argentina, Australia, and South Africa have plummeted to record lows—minus 35 degrees in some cases—bringing unseasonal blizzards that have buried towns under several metres of snow. In Southeast Asia, the monsoon season has failed. Rivers are running dry. Millions are without clean drinking water."

A new map appeared, showing swirling storm systems and pulsing heat zones.

"Satellite data in the last 24 hours has recorded multiple sudden atmospheric shifts over the Atlantic, the Mediterranean, and the Pacific—multiple earthquakes around the Pacific Rim, prompting a tsunami warning.These events do not match any recognised

natural patterns."

She then announced, *"Governments are attributing this to climate change...."*

The camera lingered on her face for a moment too long. Behind the professional mask, there was something else—worry and disbelief.

"Further updates will follow every hour. All citizens are urged to remain calm, follow official guidance, stay indoors, and prepare for ongoing disruption. This is not a temporary event. We are in uncharted territory."

The screen faded to aerial footage of parched farmland, flooded cities, and wildfire smoke curling into the stratosphere, icebergs in the Gulf of Mexico.

Many scientists and meteorologists speaking off the record described the weather as unprecedented and inexplicable. Hinting strongly that it is being engineered by man, claiming it was weather manipulation, not nature, and definitely not natural events. A war on the planet!

Robert switched off the television. The silence that followed was heavier than the heat outside.

He pulled on his boots, stepped outside, jumped into his jeep, and began the drive up to Misty Point.

MISTY POINT

It started with a sound Robert couldn't explain. A low mechanical hum, a vibration that vibrated through his body and the soles of his boots.

He stood at the edge of the cliffs. The North Sea was clawing at the rocks far below, and the clouds were hanging low over Misty Point.

Misty Point is located in Norfolk, in the village of Overstrand, close to the famous town of Cromer—a small, quaint, sleepy village where not much happened. There were around one or two shops, a small local pub, a post office, and a mysterious hotel.

Norfolk was surrounded by RAF bases and listening stations left over from the Cold War. Overstrand had a little-known weather station set deep amongst the forest, just out of the village. All you could see from the road were antennas and radio masts.

It was still and active, a covert operation. The local rumour was that it had always been run by the CIA, with two other radar stations dotted along the coastline, just outside of the village.

The locals often complained about the humming and the large 6g masts. They said it affected the weather, their health, and caused headaches, but of course, nothing was ever done about it.

Misty Point was a place that most locals avoided. There were stories from long ago of strange, unexplained noises, disappearances, cliffs collapsing, lights, and vibrations in that area. Only the occasional dog walkers or hikers ever went there. Now the area was used for sheep farming.

The government had quietly fenced off parts of the area decades ago. No reason was ever given, and no official notice. 'KEEP OFF' danger signs had been put up. Often, you would see a few of the military soldiers around the area.

Misty Point and the caves had been a well-known area to shipping and the fishing industry for hundreds of years. It was a well-used cargo shipping lane, connecting Amsterdam through the North Sea to the Atlantic. Unfortunately, it was also a place where ships often found themselves in trouble, and over the years, many sank. A few ran aground, even with an automated lighthouse.

The weather in that area was always erratic and made it difficult, even with modern technology and radar. Vicious winds and huge sea swells would often appear out of nowhere.

As the name suggests, it was often misty and covered in fog. A sailor's nightmare, though, because the rocks went far out into the sea in peaks, shaped like a dragon's tail. The local fishermen all had stories about the area—sunken villages, ships found with no crew, and sunken shipwrecks. The sensible ones never fished there anymore and gave it a wide berth.

In the calmer months, the area was often used by divers looking for anything left in the wrecks. A cargo ship full of cars and electronics sank there a few years ago. Every now and then, cars, random electronic, and motorbikes would get washed up. It was great for salvaging precious metals.

Robert was taking an early morning walk—*'thinking time,'* as he called it. He paid no attention to the slight hum and the old rumours, or the military signs to keep away. He had a history of ignoring boundaries and rules.

He spotted it wedged between two rocks and a fence post, half-covered in moss and white sand. A small, leathery, old-looking pouch. It was covered in a damp, salty slime.

He knelt down and pulled it free. He wiped it clean and unfastened the metal clasp. He noticed that there was unusually no rust on the clasp, as if it were new. He opened it to see if there was a name or an address of the owner in it.

Inside were a few items that looked unusual. It felt strange as he touched them. There was a hand-drawn map etched onto a piece of fabric. Showing a large structure, it had its own iridescent glow and shimmered faintly in the light. He then saw a brittle notebook filled with symbols that looked like hieroglyphs: an 'Ankh' symbol, strange languages, and lines that read like command codes.

Lastly, his gaze landed on a slim metal object that resembled some sort of compass, but unlike any compass he'd ever seen. As he touched it, the needles started to spin slowly, with what looked like small glass cubes at each end.

From the moment he touched the compass, it felt like it had attached itself to Robert, to his mind.

A single paragraph was inked on the last page of the notebook in shaky block letters:

'THE ATMOS CORE.
MUST NOT FALL INTO MILITARY HANDS.
IF ACTIVATED INCORRECTLY,
IT WILL COLLAPSE THE GLOBAL WEATHER SYSTEMS
WITHIN 72 HOURS.'

Robert's skin turned cold. He snapped the pouch shut and looked towards the black mouth of Misty Caves. The wind fell still for a moment. Then the hum came again—stronger, deeper, unmistakably artificial.

He took out his phone to call Bruno. Bruno was Robert's best friend from his old school.

But of course, there was no signal. *Norfolk is dreadful for phone signals,* he thought.

He turned, jogged back towards the car park, and jumped over the fence. As he got into his jeep, in the distance, there was a big 'booming sound' across the cliffs—like something massive had started to awaken, a turning deep underground. He drove back to the village to find his friend, Bruno.

THE COMPASS

Bruno had always said Robert had 'a nose for trouble.' Usually, he meant it as a joke. This time, Robert wasn't laughing.

They sat in Bruno's shed, which smelled like old oil and forgotten Red Bull. Half a dismantled drone lay on the workbench, propellers missing, wires spilling out like intestines. It was Bruno's latest project. The shed was full of stuff: a half-eaten sandwich, computer chips, old hard drives, a very cluttered workbench, an old sofa, and his dad's old rocking chair tucked into the corner, covered in monthly science magazines and old newspapers.

Bruno was a bit of a hippy, a nonconformist. He lived off-grid and was convinced the government was not to be trusted. He was a very clever man, even though he dropped out of college. Bruno was well-known to the locals. He was always up to something strange—listening in on military communications, flying drones in places that he shouldn't, and talking about conspiracy theories in the local pub. The locals just called him a little weird but harmless.

Bruno was Robert's best friend; they were almost brothers. What Robert didn't know Bruno did.

Bruno was a technical wizard; Nicole Tesla was his hero. Anything to do with electricity, computers, or coding—he was *the* man to go to. He had a little side hustle going on, fixing people's PCs and converting old video film into digital film.

He always had a project on the go; trying to create 'free energy' was his latest endeavour.

His shed was a haven for '*tech nerds.*' It was full of odd equipment, Geiger counters, computer screens, thermostatic meters, boxes full of different coloured wires and cables, bits of circuitry, old computers stacked in corners, and some very odd-looking tools. A geek's paradise.

The pouch sat between them like a loaded gun.

Robert said, "Bruno, take a look at this. I just found it, up on the cliffs."

"Mate," Bruno said, wiping his hands and eyeing the map, "this looks like a Dungeons & Dragons campaign. Are you sure someone's not trying to pull your leg, or maybe some kids just dropped it?"

"I know what a prank looks like." Robert flicked open the pouch again and pulled out the compass-like device. "This thing has a sort of very faint vibration, and it hums like it is powered. But I can't see any way to open it or any charging points. There's no power source that I can see, no microchip, nothing, but I can feel it, like it has a connection to me; it just feels strange. And it points in different directions when I touch it. It's now pointing ... there."

It was pointing towards the northeast wall of the shed. Bruno leaned back, sceptical. "What, straight through Farmer Clancy's sheep farm?" He smirked!

"No. Through to the radar station and Misty Point cave. That direction. And look at this." Robert slid over

12

the notebook.

Bruno scanned the page with the warning.

'ATMOS CORE...
MILITARY HANDS...
WEATHER MANIPULATION...
72 HOURS."

"Okay, it's either a very elaborate prank … or someone's genuinely unhinged. What possibility do you think it can be?"

Robert leaned in. "I felt something coming from the cave area, Something pulsating, something sort of alive. But not like an animal or mechanical; it was just like a presence, a feeling! The place had a low throb, a strange vibration, and a quiet hum."

Bruno smirked. "You've watched *Alien* too many times."

But Robert was not joking. He pulled out his phone and showed Bruno a photo he'd taken before losing signal. The entrance of the cave was glowing. A very faint blueish colour with an orange shimmer. Definitely not anything natural, it was artificial.

Bruno squinted. "Okay, that … isn't natural. How come no one else knows or hears it, but you did? Why was it only you?"

Robert explained, "I was just on a normal morning walk, up on the cliffs, then something caught my eye in the rocks and grass. I saw the pouch, and at that moment, I felt a strange noise in my head, like a voice, but it was only in my mind. It was as if someone or something wanted me to notice it. When I touched it, I instantly felt like I had a connection to it. A very strange feeling. It was as if it wanted me to go into the cave and find something."

"Find what?" said Bruno.

Robert said, "I do not know yet. But we must take a look—*that I know*. It feels like it's especially important. I came straight back here; I knew you would have some ideas of what it might be!"

Bruno got up and grabbed his laptop. "Ok, let me check something first."

A few keystrokes later, Bruno had three satellite maps on the screens. Three computer screens, all showing different information and graphs.

"The cave system was last officially mapped in the 70s. But check this out. This is a thermal imaging scan from last year." He pulled up a classified-looking heatmap. "This one is showing all of the Overstrand coastline from only last month!"

"This screen," he went on, "is showing a ground penetrating scan. Something big is underground there, and I don't mean a *cave*! I don't think it's a cave at all!

"The other screen is showing radiation coming from the 6G masts in the village, and further around the coastline, it's showing a lot of energy spikes. Unfortunately, most of the reports have been redacted, so I can't read the full accounts. And I'm picking something in the shed; it must be that compass thing you have. It's radiating a very low signal. And this is a satellite image of the radar station, but I can't make it out! Yet."

As Bruno looked more, he said, "I have also found an RAF manifest, dated only last week, a delivery! The railway station is going to receive a delivery of two carriages. The inventory manifest has also been redacted, so I presume the equipment.

Overstrand station was hardly used anymore. The waiting room was still intact, but the platform had nearly all been removed. It had two sidings. Some of the tracks had been pulled up. It really was fairly derelict. The only people who ever used it were the occasional airmen from the radar station, a few train spotters, and random dog walkers, but it was not a place the locals usually went to. It was quite dangerous, with all the old, rusty, corrugated metal and old railway tracks strewn all over the place; it was very overgrown.

The railway track ran along the coast. It was once a mainline train service, bringing the rich and famous to holiday in Cromer. Those days had long gone. Now, it was used very occasionally. It used to run from Cromer to South Repps, Overstrand, Mundesley, and through to Norwich. It was just a sleepy old railway line. Its days were numbered.

It was nice to see a train occasionally running through the countryside. Very picturesque. But the yearly droughts were taking their toll on the countryside.

Bruno was still looking through all the data on the screens. "There is activity there. Fluctuating Power signatures. Unusual, regular pulses. There's something like a buried grid of some sort? And whatever the cave thing is, it's *big*! But these military scans show it's nothing like a typical cave setup. In fact, it just looks like a massive structure."

Robert frowned. "How did you do all of that so quickly?"

Bruno just grinned. "They shouldn't have used the same passwords for three defence contractors, so it was easy to bypass weak passwords!"

Suddenly, Bruno's screens flickered with interference, and the pictures were lost. All the computers crashed. All the equipment in the shed went dead. Bruno jumped up and went to the fuse box. "Nothing has tripped! I have got no idea what caused that. That's very strange. Nothing has overloaded. Luckily, I have backups for everything."

A faint mechanical chirp came from Bruno's dismantled drone on the workbench. Robert and Bruno stepped back from the bench.

The drone powered itself up and slightly wiggled. It was in bits all over the workbench, and there was not even a battery installed; it was in pieces. It sort of woke up.

Bruno said, "That's just Impossible!" He froze. "That's not funny. I do not understand... It cannot happen! It has no power!"

"I didn't touch anything," said Robert

The drone's camera lenses slowly rose up and rotated a full 360 degrees, the camera slowly looking up and down the room, scanning the computer screens. A beam of light scanned the shed walls and then locked onto Robert's face. It scanned his face multiple times.

It made a whooshing noise and then just shut down as fast as it had turned up.

Bruno stood up slowly in shock. "Okay. So now you have got my attention. The drone has no power source—it's in bits. What just happened? It's crazy! It's impossible. Somehow, someone must have remotely taken control over it. Someone must have noticed me scanning the military computers and has now done some sort of feedback loop to find us. That's not possible with normal technology. Trust me. That takes very, very advanced equipment and technology to be able to do that!"

Bruno went on, "I think someone knows we're looking into things that we shouldn't be."

Bruno spent a few hours getting all of his screens and electrical stuff back online. They decided to drive back up to Misty Point to see if anything was going on.

Bruno sat in the passenger seat of Robert's old, rusty Jeep, maps spread across his lap. "You know what's ten miles from Misty Point?" he said. "RAF Lakenheath and RAF Coltishall."

RAF Lakenheath was a huge UK-American joint air base. Coltishall flew Jaguar attack fighters. It also had a large communication centre in contact with stations all over the globe.

Lakenheath was the major air base. F-22 Raptors, F-15E Strike Eagles, and F-35A Lightning II jet fighters all flew from there. A squadron of helicopters, plus in-flight refuelling tankers, was also stationed at the base.

"It is the biggest base in the UK," explained Bruno. It still stores atomic weapons and has nuclear-reinforced hangars for the jet fighters. It's an active technical, intelligence-led air force installation.

"I have heard … stories. Weird ones. About subterranean levels that go deep underground, testing new technologies and weapons since World War II, strange unexplained lights, and energy readings that no one could explain and always denied. And of course, we all know what happened at Rendlesham Forest!!!" said Bruno gleefully.

Robert stared out at the dark countryside as they drove toward the cliffs.

"The cave structures, or whatever it is, are very old. Much older than the air force base."

"Is that why the bases were constructed there, do you think?" asked Robert.

Bruno explained, "From the scans, I told you, the caves don't appear to be caves at all. There is something under the cave mouth, a structure. The layers are too uniform. It's definitely not how cave formations are."

"So why now?" he asked Robert a moment later. "Why do you think someone put that pouch where you could find it?

"I don't think it was left," Robert said. "I think it was placed by something who wanted it to be found."

The Jeep headlights swung across the mist-covered road just before the hill dipped, and they both saw them. Two black, unmarked helicopters, hovering and circling over the cliffs, near the cave mouth, as if looking for something. It swung over the Jeep, low and aggressively. Super bright lights lit up the fields and the cave mouth. Then one vanished into the fog, and the other circled around twice more, checking out the woodlands and craggy shoreline, and then sped off. It was definitely looking for something.

As the helicopters flew away, their bright lights were still lighting up the beaches and cliffs, lights darting up and down, even looking out to sea at one point.

In the distance, in the next village, they could see that the radar dome was lit up. Two miles away, a village called Mundesley had a radar dome. Not very big on the outside, but it had a large underground facility; unusually, tonight it was lit up like a Christmas tree. Something was going on. The helicopters were flying low around it. The public had been told months ago that it was closing down as it served no purpose anymore. But not tonight!

THE DOME

The climb down to the cave mouth took Robert and Bruno nearly half an hour. Loose stones rattled away under their boots, echoing endlessly into the darkness below. By the time they reached the ledge, both were slick with sweat, hands scraped raw against the jagged chalk.

The cave yawned before them, vast, silent, and faintly glowing. That strange blue light shimmered like mist caught in stone, neither natural nor artificial. Robert's compass vibrated violently in his pocket. He pulled it out. The cube-shaped device had stopped spinning. Its needle pointed firmly into the mouth of the cave. One of its smaller cubes glowed yellow.

"It wants us to go in," Robert whispered.

The air inside was heavy, muffled, as though the cave itself swallowed sound. Their footsteps crunched on gravel, but no echo came back. Only the occasional drip of unseen water broke the silence.

The walls gleamed faintly with embedded stones, not quartz, not crystal, but something else entirely. When Robert shone his torch, the strange gems caught the light, pulsed faintly, and then began glowing on their own. The cave seemed to be waking up.

"It's warm," Bruno muttered, running his hand over the wall. "Smooth. Like melted glass."

The stones pulsed faintly where he touched them, orange and green sparks flickering across their surface. The glow wasn't a reflection; it was a reaction.

"Alive," Robert murmured. "Almost like it knows we're here."

The tunnels split ahead, branching into a maze. Some openings were wide enough for vehicles, while others were tight enough to squeeze through sideways. Above certain entrances, scratches marked the walls, some of them deliberate symbols and some frantic gouges. A torn scrap of clothing dangled from a jagged edge. Empty bottles, rusted tin cans, and even bones lay scattered across the floor.

Bruno froze. "Jesus Christ ... look."

Human remains—long since dried and scattered—rested at the base of one passage. Ribs crushed. A skull with the jaw snapped clean off.

Panic flared in Bruno's eyes. "I'm getting out. Now! I don't care what that thing in your hand says."

Robert grabbed his arm. "No. We keep going."

Bruno said, "Have you lost your mind?! Look at this place—people came in here and died! They got lost, trapped, eaten—God knows what!"

Robert said in a calm voice, "Don't you remember that couple from Holt that disappeared while they were on a walk? All that was found was their dog, left abandoned on the beach. They have never been seen again. Over the years, many people have gone missing in this area."

Robert said again, "Bruno, in a cave like this, you could get lost forever, with so many tunnels and turns, and the walls looking all the same.

"It's important," Robert cut in sharply. His voice cracked with something he didn't understand himself. "Can't you feel it? This place … it matters. To all of us."

It took him fifteen long minutes to calm Bruno down. Finally, Bruno exhaled shakily. "You promise me we'll tell the police. About the bones and the missing people."

Robert nodded. "We will. But first, we find out why we were brought here."

The compass guided them deeper. Yellow for correct, and red for wrong. Each time they hesitated, the cubes pulsed more insistently, like a heartbeat urging them forward. Behind them, the walls seemed to heal, the stone slowly knitting together until the paths appeared to vanish.

"Jesus … it's sealing us in," Bruno whispered.

For over an hour, they walked, deeper and deeper, until the air changed. It grew cooler and fresher, filled with a faint ozone tang. The floor underfoot became smooth—not rock, but polished slabs fitted with microscopic precision.

And then the tunnel opened. They stepped into a space so vast their torches were useless. A dome, fifty feet high at its lowest point, spanning wider than a football stadium. The walls shimmered with a mirrored polish, as seamless and smooth as a moulded jelly. No cracks, no joins, no evidence of tools. Just flawless, an unearthly design.

Light came from everywhere and nowhere, flooding the chamber in a clinical glow. Strange glyphs pulsed across the walls in flowing streams—hieroglyphs, geometric shapes, star constellations. At times, they looked like writing; at other times, like shifting code.

At the centre of the dome, an immense egg-shaped structure hovered a few feet off the ground. Twice the size of two articulated lorries side by side, its surface shimmering like liquid metal. Around it, towering crystal obelisks rose to the ceiling, suspended in the air as though gravity had been dismissed.

They hummed faintly, resonating with Robert's compass, which now glowed so brightly he could barely hold it.

Bruno whispered, "That…that's no cave. This isn't even Earth anymore."

Robert walked forward, drawn by a pull he couldn't resist. As he touched the surface of the egg, colour rippled across it, looking like liquid metal. As his fingers touched it, colours moved to where his fingers were.

Robert exclaimed, "Bruno, this machine … it's a symbiotic entity. It's responding to my mind and my touch; it's as if it's alive!

And then it spoke. Not aloud. In his head. Giving Robert information and, more importantly, instructions.

"Wow, I've read about this technology," said Bruno. "A **symbiotic** entity."

Robert gasped. "It's alive. It's not a machine—it's a life form."

The walls flared. Images erupted across the dome's surface like a planetarium gone mad, like Earth, as seen from space. The blue ball normally looks delicate, vibrant, and alone in space, but now, it looked gray and angry as furious storms ripped across continents.

Hurricanes, tornadoes, and entire cities were drowning. Crops withered to dust. Deserts spread. Huge fires with smoke plumes high in the stratosphere. People moved like herds, millions fleeing starvation. The planet looked devastated.

Then a change. The images softened, showing another version of Earth. Balanced. Beautiful blue skies, oceans stable, ice cap still intact, fields of crops lush and golden, the Amazon Forests unbroken, even extinct animals roaming free, and rivers full and clean.

Words scrolled beneath in an alien script, translating into English directly in Robert's mind:

> *"Atmospheric Regulation System*
> *— Protocol Failure Imminent.*
> *All systems must be reset."*

Bruno's face had gone white. "Robert … are you trying to tell me that it's controlling the weather?

"No," Robert's voice shook. "Not controlling. Balancing. It was built to protect us. But something's tampered with it, and I think that there is more than just this one."

The compass pulsed violently, flashing red and

yellow.

The voice returned in Robert's mind:

"Equilibrium engine, calibrated for human and mammal survival."

Robert's heart pounded. "This was built for us or a civilisation like us! It was built to make the planet thrive, but it's been tampered with and is not functioning properly. We have to reset it somehow! We will all watch the world unravel without it working correctly," Robert said grimly, "and I think the government and military are behind it."

"Human interference detected.
Stabilisation incomplete.
Reset required now!"

Suddenly, the dome darkened. The glowing glyphs flared blood red. A warning appeared...

"Unauthorised Access Detected.
Signal Intercepted.
Counter-Infiltration Imminent."

The compass screamed in Robert's hand, vibrating red, then flashing yellow—pointing straight at a section of the wall. With a thunderous whoosh, a doorway materialised.

"Run!" Robert yelled.

Light spilled across the floor in a glowing path, leading directly into the opening. Behind them, the dome shuddered. The egg pulsed with crimson light.

And then—faint, echoing through Robert's skull— the voice whispered one final command:

"Hurry."

WEATHER WARFARE

On the edge of the Norfolk countryside, beneath a nondescript hangar at the Coltishall base, a red light started to blink on a command console.

Then all of a sudden, other monitors started to get alerts, and technicians rushed to their workstations. All started to receive anomalous readings.

All the technicians started reporting movements on their screens. All stations were now looking at the monitors, and motion detectors and heat monitors had all been activated.

A technician tapped the screen, looking confused. "Sir," he called, "we've just picked up a thermal surge, and motion detectors have been activated. And we have automatically gone to Red alert? SIR. All the motion sensors from the eastern caves have gone crazy, Sir, all have been activated.

"SIR. It's a foreign signature, not Russian or Chinese. It's unlike any signal we have received before; it's right at the end of the spectrum, a very low vibration. It seems to have biometric signatures. SIR!"

The colonel stepped forward in his dark, crisp uniform, eyes colder than steel. The colonel knew what this could mean and how important the ramifications could be. It was the worst news, and he was expecting it.

Since the end of WW2, the military had been developing equipment to manipulate the weather. Geoengineering, a way to create weather wherever they wanted. They could manipulate droughts, severe flooding, extreme rainfall, temperatures—one day 45 °C, the next -2 °C—and even wildfires could be created.

'CEM' trails from the planes seeded the skies with poison, which affected the soil and the water system, and were designed to reflect the sun 'to cool the planet.'

The military had nearly taken all control.

The nickname they gave it was 'climate change' and 'global warming,' and then blamed it on the population. Overpopulation and burning too much fossil fuel, when in reality, it was the military. It was really weather manipulation; it was control!

The military also knew there was an ancient system in place to balance the globe's weather. They did not fully understand where it was and how it worked, but they were getting close. The military operation was to find it and destroy it.

The listening post at Overstrand was part of the mission that only the NSA, CIA, and a few other top military brass knew about. It was a part of the *top-secret* operation, the military's attempt to try and control the weather patterns around the world.

The CIA had recovered a piece of Nikola Tesla tech, back in the 40s, that seemed to have an effect and started to alter the ionosphere. The CIA and NSA were trying to understand how it worked. The scientist understood what it did, but quite how.

THE CIA had built a laboratory in Alaska to build and hide the experimental system. The Americans called it *HAARP*, an ionosphere heater that moves the atmosphere up and down. A nice sounding piece of equipment, but when working efficiently, it could actually move weather patterns, causing hurricanes, disrupt the Earth's mantle, causing earthquakes, which would produce chaos and disturb the weather systems everywhere. It can also create massive wildfires, which we have seen all over the world. It's a weapon, hidden in plain sight!

Up to this point, they were only having limited success, but were still changing the weather! Around the world, every country had reported extreme weather, be it flooding, extreme heat, or flash flooding. The military wanted to control the weather for its own ends, and not necessarily for the good of all mankind.

The station commander shouted, "Send in the 'reconnaissance drones.' And prep an extraction crew. NOW!"

"But, Sir. The readout suggests … it's stabilising the weather. It is becoming beneficial. The weather systems are calming, and storms over South Africa are stabilising. Sir."

Another operative called out. It had also detected movement. Three heat signatures.

The colonel stared at his subordinates. "This is classified, top secret. Nothing leaves this control centre. It's all on a need-to-know basis." He continued to look at the screen.

"Send in the drones and a tactical team right now!" the colonel ordered. "No one must leave that cave. No press, no TV. Nothing leaves this room. Everyone must report to me directly, and me only! In fact, I want it totally sealed and locked down. Now!" he shouted.

WEATHER MODIFICATION

The compass in Robert's hand buzzed wildly. The crystal dimmed. A final projection flickered into place.

Another message appeared.

This time, a sort of face came into view. Not human exactly, but humanoid—the features smooth, elongated, no hair, ageless.

It spoke, but the voice went directly into Robert's mind. It seemed to be everywhere.

"You have found this chamber. Our system is failing. You must reach the main core. It lies beneath this level! Your military and governments, they're trying to use a system they call HAARP to destabilise the ionosphere. Changing the weather around the world and using it as a weapon. Seeding the skies with chemicals to create more clouds to dim the sunlight. Purposely poisoning the water systems and the soil. Making people leave their homes and putting pressure on other countries to house them. This is

your military's attempt to destabilise other countries with overpopulation, creating chaos."

The voice continued:

"This dome is not just a ship! It is the keystone to the systems we put in place. This is part of a 'moon' system we put in place centuries ago. The climate will not hold much longer without them all working as one."

Robert frowned. "It told me we must get to the core; it's even deeper. This is just the beginning."

Bruno nodded, swallowing hard. "We're in the middle of something massive. And we are not the only ones looking."

From above, a sudden mechanical whizzing sound echoed through the chamber.

A black dot zipped across the ceiling—a surveillance drone—its infrared laser scanning the crystal obelisk and the engine. Infrared lights scanned the dome and structures. It then picked up Robert and Bruno and zoomed off into the darkness.

"We're definitely not alone now; the military has found us!" Bruno picked up the compass, flashing yellow. "We've got to move. This way. Now!

At that moment, Robert heard a noise from behind one of the crystal columns.

A girl stood up, frantically waving her hands and arms into the air.

Robert said, "Who are you, and where the hell did you come from? How long have you been here? In fact, how did you find this?"

She said, "My name is Kate. Please, I don't mean any harm. I followed you. I knew something was weird in this area, and something strange was going on. I have heard stories regarding these caves. I just wanted to be involved. I needed to know! I can help, I feel it. I have seen the helicopters over the cliffs, the last few nights. Last night, the helicopters were back. That's when I saw you two heading down into the cliffs, so I followed you into the cave system. I felt it, like you did. I was being drawn in. Something was calling me."

Robert said, "Can you hear it, too?"

Kate said, "I can't hear it, but I can feel it's a living entity, and I can feel its presence calling me."

Bruno said, "I think I recognise you from a science

class at the university and a protest group that stormed Cromer council offices last summer. *'WOMEN AGAINST CLIMATE CHANGE'* I think..."

Kate said, "I know what's going on. I want to help. I have heard what you have been saying, and I need to help you stop whatever is going wrong."

Kate also told the boys that the military had started to arrive in force, up on the top of Misty Point.

Robert said, "Come on, we have got to move."

: The mothership connection

The drone had gone. They now knew someone would not be far behind them.

Robert gripped the compass, still pulsing yellow. The needles still pointed towards the doorway, a narrow passage at the far end of the dome.

"That drone must be linked to Lakenheath," Bruno said.

Kate said, "It's not just Lakenheath; it's also RAF Coltishall that's involved with this. RAF Coltishall is in charge of special global intelligence operations. On the eastern sea front, they monitor all sorts of stuff from around the globe. I picked up one of their transmission on my CB radio. A lot of chatter on the airwaves lately."

"They'll want this shut down," Robert replied. "Or worse, destroyed."

There was no turning back now.

The descent began.

The corridor narrowed. The compass flashed yellow, the walls resembling tubing.

Like an umbilical cord leading down from the large dome. More energetic patterns followed them along the walls—spirals, geometric codes, etched with surgical precision. The light danced across the gemstones.

"Looks like … energy routing," Bruno murmured. "Some kind of resonance grid. Tesla talked about stuff like this."

Robert nodded, scanning the markings. "They're encrypted. But the symbols are repeating. They are instructions."

The passage opened into an even bigger circular dome. This dome was filled with more pillars of glowing quartz. Hundreds of pillars rose from the floor to the ceiling, like skeletal remains of an ancient temple. At the centre, a tube descended into more chambers.

"This is a space vehicle; it's massive," said Bruno. "It must have been here for centuries."

Robert received another massage. He approached a terminal and gently placed his hand on it. It powered up instantly. "Biometric, symbiotic recognition," he muttered. "It's amazing. It's an intelligent design centuries ahead of anything we have or even thought of."

Lines of code scrolled rapidly. Then something appeared on-screen:

"System Override Detected. Atmospheric Collapse Protocol – 67% Complete."

Robert said, "It's communicating with me; it told me what I must do. We must act fast." He touched the wall, and certain parts lit up, like in a sequence. "The huge crystals are the power source; that's why they are starting to light up and vibrate," said Robert. "I've done it. It's nearly at full power and running correctly. I have one more thing to do. I have to align these smaller crystals."

Bruno cursed. "They've already started trying to mess with the systems."

"But not this one. This is the mothership. There must be more!"

***

Lakenheath Control Centre

Colonel JC McGowan stood in the war room, the soft glow of a dozen wall-sized screens painting his face in flickers of blue and red. The screens pulsed with global data: swirling cloud formations, rainfall projections, and graphs of collapsing harvests.

On one screen, Africa lit up in orange and crimson—zones of drought, famine, and rapidly failing crops. Another screen showed monsoons hammering Asia, rivers breaking their banks, cities drowning. Entire continents blinked with warning lights, as if the Earth itself were crying out.

McGowan folded his hands behind his back, the medals on his chest catching the sterile light. His voice, low and deliberate, carried across the hushed chamber.

"Chaos, mass migration brings control," he said.

Around the polished table sat a dozen of the world's most powerful figures—presidents, generals, ministers—each cloaked in shadow, their faces illuminated only when they leaned towards the screens. None interrupted.

"Starvation. Flooding. Collapse of order," McGowan continued. "The people will beg for structure. They will beg for help. And when that moment comes, when the world is on its knees, we will be ready." He turned to the screen behind him, where a schematic of the weather-control array pulsed like a beating heart.

"If we offer the solution, we own the outcome. We will control water, energy, and food. Every harvest, every mouthful of rice, every drop of clean water. Their movements, their money, and their survival itself will flow through our hands. The old systems will die. And in their place ... a new world order. The elite will rain!"

The words hung heavy in the chamber. Some of the men nodded slowly. Others sat rigid, the enormity of the plan dawning on them.

A general, his uniform thick with ribbons, cleared his throat. "And the intruders?" he asked. "What of the ones who found the caves? The machine?"

McGowan's lips curled into the faintest of smiles. "They are nothing," he said. "Children playing with shadows. My troops are already there. If they come within reach of the array, they will be … neutralised."

His gaze hardened, and for a brief moment, the hum of machinery seemed to deepen, as though the very room shared his menace.

"No one," McGowan said softly, "will stand in the way of destiny."

SPACESHIP MOON

Robert touched the control switches on a side console. It immediately lit up. The voice in his mind gave him the final set of instructions.

Robert opened the hatch, and it slid out. It was filled with multi coloured crystal rods about nine inches long, in rows. The crystals looked like a jigsaw puzzle, but *vertical*.

The voice was telling him how to rearrange the crystals into the correct sequence to bring everything back online. Blue, orange, red, and black crystals—he slowly finished it. Immediately, the hatch slid shut, and out came a rhythmic sound of something powering up.

"What's that?" Bruno asked. "If we can extract it, we could power an entire city. This isn't just a weather stabiliser—t's a vault of technologies. Just imagine if we had this technology, what we could do with it. A gift for humanity."

"No," said Robert, "it is a *gift* for the world. It is our weather system."

The system flared to life. It was now fully activated. With the two minds connected, the crystals and dome started to move. Lights rippled across the ceiling. Two floating spheres began to rotate, faster and faster.

The whole room looked like it was moving, spinning, and pulsating. A map of Earth appeared on screens around the dome. The screens showed what was happening to the countries around the world in *real* time.

Robert stated, "This machine is working in conjunction with the Moon! We have done it; it is now fully back online. Its connections to the moon are fully restored."

Bruno and Katie looked puzzled.

"The Moon, what's the Moon got to do with all this?"

Robert started to explain. "Listen," he said in a very stern voice. "We don't have much time, but please try to understand." He proceeded to tell Katie and Bruno exactly what was happening, as he tried to understand it himself. "The Moon is artificial. It was placed there many centuries ago. It controls the weather systems while working with the machines. Before the moon was put in place, the world's weather was not stable, and the Earth had a wobble. The Earth was covered in a fine watery mist that blocked 70% of the sunlight. The Earth was green, but there wasn't enough sunlight for bigger animals to thrive.

"With the Moon in place and the machines working together, the tides were created. When the Moon was rolled into place, the watery mist fell. All the water fell from the sky, creating one of the great floods. All the civilisations from around the globe had their own flood stories, " like the Noah's ark story from the Bible." The ark was a DNA bank with all the species of the Earth stored inside. The next reset!

"We are now on reset no six. We are not the first civilisation! There have been many before us. The

Atlanteans were the fifth experiment. Atlantis is a sunken civilisation. Because of its great flood, it reset. We are much older than we are told in our history books. All the other experiments were failures! So we were created and became the sixth generation … or experiment. The Moon helped create our planet's unique weather systems. It stabilised our weather, which we enjoy today—summer, winter, and Autumn. Only possible because of the Moon and the alien machines, Earth was given its unique tilt, which creates our seasons.

"The Moon affects the tides, forcing the seas to spread nutrients around the world. Animals could grow with more sunlight, and the dinosaurs, whales, and sea life all started to thrive. There are crops and clean rivers in abundance. Oxygen and CO2 are at levels that humans and mammals need them to be, with normal sea levels, and the North and South poles as they should be. The trade winds blew. The world's weather system became calmer, more predictable, and had a vibrant, thriving ecosystem, exactly right for the animals and humans.

"The Moon had been placed in orbit, into our sky, centuries ago by an alien race, *the Anunnaki*, from a planet called Nibiru. NASA refers to it as Planet 9. Working with the machines, they set and balanced our weather system. Maybe Earth's Guardians! Or our creators," Robert said. "Nibiru is coming back, its orbit should bring it back in 2051!!"

Bruno said, "Are you taking the piss! How can that be? Of course, the Moon is a natural satellite."

Robert said, "Listen to me, please. It's important that you really understand."

MOON MISSION

"The Apollo missions. That was the excuse to send men to the Moon. The Apollo missions in the sixties and seventies were sent to understand what the Moon really was, how it could be so big compared to other moons, and if a civilisation had been there, or still was there! The Americans also needed to beat Russia to the Moon for political reasons. The Russian scientists knew the Moon was not what we were told or thought it was. NASA knew from scans that the moon was hollow, and the surface was made of a type of super-strong, unknown metal. Its orbit is unlike any other moon in our solar system.

"The most efficient shape for a spacecraft is a sphere. The metal on the Moon is super strong and perfect for building a space station. It's an artificial satellite. The Moon's crust is thicker and more dense than the subterranean rock, which is also strange. It's as if the crust of the Moon is acting like a shield. Strange lights have been seen coming from the Moon over the centuries, particularly a blue light that appears every now and then. The Moon has no iron core, but its rocks are magnetic. The Moon's orbit is nearly perfectly circular. And it is the only moon to be like this in our solar system. The Moon is too big and too close to the Earth to just be there, without being placed. No one knows how, but it was done!

"NASA knew someone or something had been on the Moon for years; of course, the military always took it as a threat, thinking it was the Chinese or Russians. There are ancient stories from the Zulu tribes. The Zulu tribes talk about times before the moon was there. There are legends that say that the Moon was rolled into place by a dragon, and they talk about it as an egg, hollow. The moon stabilised our weather.

"In the Grand Canyon area, the Native, Navajo, and Hopi tribes talk about star beings from around the globe. The indigenous people all talk about travellers from distance places and have legends of when the moon wasn't there. The star children. The ant people, with big dark oval eyes, tall and thin with long arms, who came from the stars in flying machines that had fire. They arrived when the Moon came. So maybe all this was a gift from the gods. Alien gods! The Moon is some sort of outpost for a distance Alien race, monitoring their experiment and planet Earth.

"The astronauts from Apollo 11 left a metal disk, containing messages of peace and goodwill from all nations on Earth. Who for? They haven't gone back because the astronauts saw something very unsettling and reported they were being watched from a crater by three disk-shaped objects. NASA was concerned, but they also knew something was up there. So, it has been kept secret ever since," Robert finished.

"So you're saying the Moon landings were all faked after that?" asked Bruno.

"No," said Kate. "The government has had contact with many different visitors over the years, with some sort of agreement in place, sharing of technologies, etc. Now you know why we haven't rushed back to the

Moon!" she told them.

Robert said, "Now you know! The military is trying to control the weather algorithm. The military is manually trying to destabilise and move the jet streams from the surface. They are seeding the clouds with poisons and trying to dim the sunlight and control where it rains or does not. Causing chaos. The excuse the military gives is that it's global warming or climate change, and that it's all man-made. It's not. It's the military messing with the weather systems. All part of the 'Great Reset.'"

Bruno said, "What are you talking about? Reset!"

Robert said, "We don't have time now, but I will explain later. We have to get out now."

Bruno said, "The Moon, the Great Reset, it's all crazy talk. I love a conspiracy, but this is just ridiculous."

Robert said, "Just trust me, I will explain everything!"

Just then, the dome shook. Explosions echoed from above. Debris rained down. Dust filled the air.

"This machine is working in conjunction with the Moon! We have done it; it is now fully back online. Its connections to the moon are fully restored," said Robert.

"They're coming. They're trying to seal us in!" screamed Kate.

The dome was fully active; it felt like something big was about to happen.

The compass reared back to life, bright yellow, pointing Robert, Kate, and Bruno into a new direction.

Robert said, "It is giving us the way out! We must get out now. We must get out! This is going to take off, or worse."

They started to run down more tubes. Glancing back at the core, they could see the dome was lit up, with bright rotating colours like a tornado, and it looked as bright as the sun. It was now fully activated.

EXODUS

Outside the cave, the hills looked unchanged, apart from the military personnel and equipment.

They were out. The three quickly ran for cover. Soldiers, mobile radar trucks, and lorries were all over the place, and helicopters were in the air.

Sheep grazed quietly, and mist hovered as it always did. Then the earth beneath the caves started to rumble, and shake, moving and erupting like an earthquake. The noise was so loud, with cracking and booming sounds, that the sheep and birds started to panic and flee the area. Sheep began jumping fences, going wild and crazy in attempts to get away; some got hurt on the fences. Flocks of birds moved in thick patterns in the sky.

Robert and Bruno ran to the tree line. Kate hid behind a huge pile of logs.

Suddenly, with one more final cracking sound, the earth shook. Then it split apart, throwing soil and rocks into the sky, trees splitting and falling, and smoke/steam covered the massive ship as it slowly moved from its resting place. Pushing through the soil and roots, it then burst into the sky, making cracking sounds like static and lightning.

It hovered, just for a moment, and then it was gone. In a flash, it was just a dot in the Milky Way.

Helicopters were in the air, and soldiers on the ground. But it was too late. The whole structure had gone. A massive pit left in the ground, with small pockets of smoke rising into the sky. A scene of total devastation. The military and its plans were totally ruined.

Robert whispered, "While it was hovering, it sent me another message! We have successfully turned it back on, but there are more that we need to find and stabilise. They still need our help! Let's get back to your shed."

Bruno looked up. "Really? More?"

"Yes, Egypt. Also, somewhere in the Australian outback, a place called Pine Gap. And an underground base and settlement in the Furnace Flats in the Grand Canyon!

SHED QUARTERS

Bruno had all his computer screens on, and he and Kate were on the dark web, searching and pulling out every classified document they could find. Newspaper clippings, underground maps, and 'synthetic aperture radar' (SAR—a radar system used by satellites), looking for unusual seismic readings from around the globe.

One article stood out. A redacted MOD report from 1992. Coltishall RAF base had been picking up and tracking 'anomalous electromagnetic interference' in the ionosphere from other sites around the globe, particularly from the Cairo region in Egypt and the Grand Canyon.

Bruno explained, "By the look of these scans, there are secrets to be found, hidden underground in the canyon area, and there is a military presence in Luxor. I'm getting readings from both places, but not much about Australia. I will dig deeper. The Grand Canyon seems highly active in these scans and satellite images. I'm getting abnormal readings from the glaciers in Alaska."

Bruno said, "That's where the ice melts from the glaciers and feeds the once great Colorado River. That's what fills the Black Canyon, Lake Mead. The water levels in Lake Mead have been dropping for years. The Hoover Dam supplies hydro power to most of Nevada and Arizona, and it's already at critically low levels!"

Bruno went on, "Eventually, the Hoover Dam water levels will be so low, it will not be able to produce any electricity, and the whole structure will collapse and implode in on itself. Then millions of people and animals will be on the move, or worse, dying. Las Vegas and all the other cities and farmlands will become ghost towns very quickly!"

Robert said, "It's the heat and lack of rainfall. This is what the machines were put here for: to stop this from happening and keep a climate balance."

Robert then stated, "For some unknown reason, there are places in the Grand Canyon where no one is allowed; they are protected by the government and patrolled by helicopters. In the sixties, huge cave habitats and underground buildings were discovered with what looked like Egyptian hieroglyphics, artefacts, and a statue that looked like Buda. Once the Smithsonian museum found out about it, it was immediately closed off to the public, and anything found was quickly hidden, and the public was locked out. *What are they hiding?*"

Kate jumped up, "I have been studying the images from Giza. I know all about Egypt, the temples at Karnak, Luxor, and Cairo. The Valley Temple of Khafre is part of the Giza Pyramid Complex. I can be of help! I have friends who live on the Nile River, and I have spent time there. I can get us around without being noticed. The 'military' has been trying to reprogram and understand their machine for decades."

Robert whispered. "But they need something they do not have. Me! Plus, they don't know where the other machines are, but know they're out there. So they will be searching like crazy to find them quickly, and to try to stop us. It needs us. It needs *me*. It needs me to be the interface; the military doesn't have anything like me!"

Bruno traced a frequency to the hangar at Coltishall airbase. An unlisted building, camouflaged on satellite images. Bruno explained, "It looks like they have started the search; even the *Cheyenne Mountain Space Force Station* in Colorado is involved this time. So you know the CIA is still involved."

HANGAR 7

Hangar 7 was a cavern of shadows and steel, a space that hummed with secrets. Built into the farthest corner of RAF Coltishall, it did not appear on any map, nor did its power usage or personnel rosters ever reach official records. Those who worked here joked grimly that you didn't get posted to Hangar 7—you disappeared into it.

Commander JC McGowan stood in front of the vast, twisted wreckage of alien technology, its surfaces scarred from a crash decades earlier, with blackened plates of metal curved at impossible angles, still humming faintly with residual energy. Strange markings pulsed intermittently, as though the machine breathed. Despite decades of study, it remained largely a mystery.

The hangar smelled of oil, ozone, and the faint copper tang of scorched circuitry. A low vibration filled the space—not from any human machinery, but from the tech itself. It was alive in ways the engineers still couldn't explain. The room was full of men in white coats, with vague looks on their faces.

This place had a purpose. Hangar 7 had been reserved solely for back-engineering alien or 'exotic' equipment. Over the years, McGowan had overseen the arrival of strange devices pulled from deserts, oceans, and mountains across the world. Some had been recovered after crashes. Others, whispered in closed corridors, had been traded—offered to governments in return for something darker, something never spoken about outside protected, secure rooms.

His orders this time were absolute:

"Get it working. Reclaim control of the environment. Force chaos upon the nations, so the citizens would give in to a new world order!"

McGowan drew a cigarette from a battered packet and struck a match. The flare of orange light briefly illuminated the tired lines etched into his face. He inhaled sharply, then exhaled a slow, grey cloud that curled upwards, mixing with the hum of alien energy. A chain smoker, he always burned through more when the weight of orders pressed down hardest.

For just a moment—the briefest flicker of humanity—McGowan remembered himself as a young airman. Back then, he'd believed in service, in protection, in defending ordinary people. Now he found himself standing in front of something not of this Earth, preparing to unleash storms, droughts, and famine on the same civilians he'd once sworn to safeguard.

He crushed the thought as ruthlessly as he ground out his cigarette. Orders were orders. Turning, he issued his command. His voice was steady, cold, and without hesitation. "The three involved are now persons of interest to both the American and British governments. The use of extreme force has been authorised."

The order was transmitted instantly, encrypted signals streaking across satellites. Within minutes, it reached every listening station, airbases, and covert detachment in Europe, the US, and beyond.

Special operations units would already be preparing dossiers, facial scans, patrol lists, and putting together snatch and grab teams.

Robert Marshall. Bruno Williams. Kate Bishop Persons of interest! Three young people—now classified as enemies of the state. Their names flagged in every military database, their photographs pinned to briefing boards, their movements tracked with satellites and surveillance drones. The search was on!

A machine within the CIA's cyber-intelligence division chimed as their details were uploaded to global watchlists. Interpol red notices were drafted before the hour was out.

In a single breath, they had become fugitives. Wanted across borders. Targets with no safe haven.

McGowan lit another cigarette, his hands steady. His conscience whispered once more—*is this necessary?*—but he shut the thoughts down.

"No expense will be spared," he muttered under his breath, "to bring them in."

FUGITIVES

Robert, Bruno, and Kate had already left their homes and were now on the run.

The three of them had to find a way out of the country fast, without being seen, so airports were out of the question. They were on a mission.

Kate said she had an uncle, Sid. He worked at Felixstowe docks. Felixstowe container port is the UK's biggest shipping and container port. Sid controlled and managed all the logistics and sorting of all the containers at the port, so he knew where everything was going and where it had come from. Kate said she was confident he would help them. He was very close to Kate's family.

Robert said, "You had better contact him."

They sped off in the early hours, hoping the darkness would help them make their getaway easier. It was only 50 miles to the docks, and they used the back roads to try not to be spotted. They headed towards Felixstowe in hopes that Sid would help. Sid really was their only hope to get away from the UK and to get to Egypt.

Kate made a call, and Sid eventually, but reluctantly, agreed to help. It's not something a port manager would ever do—helping them would risk his job if found out.

Sid was best friends with one of the captains of a cargo ship that was headed for the port of Alexandria. Alexandria is a city in Egypt.

Sid asked, "Is that ok?" It was the only ship he could offer, and it was the only ship going in that direction this month.

Sid explained, "The captain would help them as much as he could, no questions asked."

"From Alexandria, it is still a long way to Cairo," said the captain.

Robert said, "Yes," and thanked Sid. "We will do the rest!"

The captain's instructions were very clear: they had to stay in their cabin.

The small cabin was in the lowest part of the ship, well away from the crew quarters, in an area that was just used as storage and old, unused equipment.

If they were spotted by the crew, all hell would break loose. Technically, they were stowaways. If caught, they would be arrested and taken to the first port. The crew would make sure of that! And it would cost the captain his job and cost the shipping company thousands in fines. The captain made it clear he would deny any knowledge of them if caught.

It took several weeks on the ship to get there. The captain brought them food and water each day, along with updates on the news.

Robert was seasick for most of the journey. He lay in the top bunk in the cramped cabin of the ship, the hum of the engines a distant murmur beneath the tension in his chest. The vessel had slipped away from the English coast under the cover of darkness, heading toward the Mediterranean—a journey cloaked in secrecy.

He had received his mission, a task that could change everything. But before proceeding, there was one crucial step—he needed to contact Helen James, the MP he barely knew but trusted. She also knew what the military was up to.

Robert activated the MP's private number, his voice low and steady. "MP James? It's Robert. We managed to get a ride on a container ship!"

"Robert, you're on a ship?" Helen's voice was calm but edged with urgency. "I wasn't sure you'd get out of the country."

"Yes. The ship left days ago. The entity gave us new information for the next phase: Egypt. That's where we're headed. It's the source of the second ancient machine controlling the weather," he said, but we still have to find it.

Helen's breath hitched slightly. "I know what you're facing."

Robert nodded grimly. "They want to rewrite the climate cycle using ancient technology. If they succeed, it will destroy the natural balance—millions will die and millions more will suffer."

"I've been working to expose their plans," she said. "I have initiated inquiries, leaks to trusted journalists—but it's a shadow war. You and your two friends are the key. Only you can find and stop what they are trying to do on the ground."

Robert said quietly. "We're counting on your help to keep the mission secret and protect us from interference."

The MP's voice hardened with resolve. "I'm your lifeline back home. No one else will know where you're going or what you're doing. I'll block every attempt to sabotage your task. We know the risks," Helen said. "But together, we stand a chance."

The line went dead.

Bruno was busy, searching for clues every day about where the machine might be. He had also started to inform internet hackers and bloggers of what the military had been up to, but nothing about the task they were embarking on.

Bruno had found two possibilities for where to start their search. Zahraa city or Abu Gorab both had unusual signals coming from the area.

Robert said, "In your opinion, which one should we go to first?"

Bruno sighed. "I think it's got to be Abu Gorab. In that area, it has the strongest anomalies underground showing on my scans, but it could be anything. It's not clear; it's very near the Great Pyramid complex. I think that's the best place to start."

With Bruno's hacking skills and Robert's connection to the core, they uploaded proof—raw video, energy readings, declassified maps—to the dark web. The truth started to spill out.

The ancient weather modification machine was real. The world was starting to wake up to what the military and elites were doing.

Even the world-famous radio presenter, Art Bell's radio show, *Coast to Coast*, had a whole program dedicated to this phenomenon. People were calling in from all over the world with stories regarding con trails in the sky, crazy, wild fires, and government conspiracies. Conspiracy forums exploded. Journalists dug into old MOD reports.

We're not alone. Read the headlines.

MP Mrs Helen James, Chair of the Environmental Ethics Committee, publicly called for more inquiries. She was the sort of woman who got things done. She would leave no stone unturned in her pursuit of the truth, a very determined powerhouse.

In reality, nothing would ever happen. The military would just hush it all up, as usual, or start an enquiry that would last for years. Nothing ever changed.

The MP had many run-ins with the government and the military over the years, but this time, she would not let it go.

The MOD and CIA were desperately trying to stop the conversations on the Internet and the dark web. More and more resources were put in place to find the trio. The world started to listen.

And Lakenheath responded. The military PR machine went into overdrive. The military denied everything, saying it was just a group of kids, illegally breaking into the facilities and damaging equipment, painting *'USA Out'* slogans. The usual petty stuff.

A warrant for their arrest had been issued, and the usual excuse and denial of anything important was happening.

THE SPHINX

A few weeks later, Robert and Kate sat in a quiet café in an Abu Gorab back street.

Bruno said, "This heat, it's unbearable. I've got sand everywhere, and my laptop is suffering in this heat."

Egypt was suffering one of its hottest springs ever; some days, it was over 55 degrees, with no rain in months. The Grand Oasis was nearly dried up. That had only happened twice before.

Adu Gorab was a dusty old town, like time had forgotten it. It was a strange place, full of markets, people hustling and shouting, trying to sell their produce, donkeys, and camels, pulling carts about, and very few cars or trucks. The air was full of the smell of something spicy; it was just an old trading town on the way to Cairo. Luckily, there were only a few Westerners, so they felt safe, but they still had to be very cautious. The CIA had ears and eyes everywhere.

Bruno held a new map in his hand. "You really think it is here?" said Bruno.

Robert nodded, "Yes, I'm getting that feeling again. The voice is telling me we need to find this monument. The *Sphinx*—I think it's inside it... We have to look for this symbol!" He drew a picture of the symbol in the sand.

Kate said, "I know that symbol! I've seen it before. It's beneath the left paw of the Sphinx."

The compass started to glow yellow. "It is in this direction," said Robert.

Kate said, "It's on the Giza Plato."

Bruno checked the map, "Yes, it's 10 miles away. We will have to get supplies, water, and transport. I will go and see what I can find in town."

Katie explained to Robert, "The Sphinx was a mythical creature with origins in ancient Egypt. At least three tunnels or vertical shafts have been found beneath the Sphinx, which lead to two large chambers, one being the fabled hall of knowledge, and another one 200 feet below the surface. It's believed to be an underground city. The Sphinx represents royalty, power, and protection, particularly associated with the Sun God Ra.

"The rumors are, beneath the paw is the 'Hall of Records.' It's been a rumor since the 30s when the Germans were searching and stealing all the religious and ancient artifacts. The top archaeologists now believe and understand that the Egyptians did not build the pyramids or the Sphinx, and that the pyramids were some sort of power station! They are so much older. You can even see the water damage on the Sphinx," she said, "but not on the pyramids. So we know the Sphinx is so much older. The Egyptians were not the builders of the pyramids, but merely the caretakers!!

"No one has found the Hall of Records yet, but scholars believe it's just there waiting to be found. The Egyptian government will not let anyone excavate. It's totally sealed, off limits. The guy in charge of ancient equities stopped most of the searches around the Sphinx, afraid of what they might find—ancient knowledge, advanced technologies, or even something that could force us to rewrite human histories, or both," Kate explained.

"A system buried beneath one of the oldest monuments on Earth," Bruno whispered. "Another ship? Another stabiliser?"

Robert frowned. "We need a plan. How can we get to the Sphinx and get in without being caught? The military won't be far behind us; they might even be in the area. We must be careful and try to get in and out as fast as possible. At least this time, we know what we're looking for and how the machines work."

Bruno was back. He had been successful in getting clothing, supplies, ropes, and most importantly, a *guide* about who had his own truck.

Akil was his name.

Robert said to Bruno, "Are you sure about this guy?"

Bruno said, "Yes. He's given me a price, and we shook on it, so we have a deal!"

Then Bruno said, "Akil told me he would take us

there for 15000 pounds. (In UK money, that's only £200.) He said, '*No trouble with police or tribesmen. I'll get you there safe; trust me.*'"

Robert's compass was glowing, and the voice was saying, '*Hurry!*'

Kate said, There is a small entrance at the back of the Sphinx. But from my memory and what I was told, it is a long, narrow, near-vertical tunnel. There is another way in, a hole under its headdress, but that's a very steep and dangerous drop; it's a long way down and with hidden traps. It's the way the relic hunters and archaeologists tried to get in. We will have more luck if we go under the cover of darkness, so we will need to leave now and make sure the torch works.

Far beneath their feet, under the sand and stone, a familiar pulse began to vibrate.

Robert stopped in his tracks. He said, "Chaps, I don't think we're just looking for a machine this time! The voice is talking about a gate? A gate of some sort, and something about a Ziggurat, but I can't understand what it means. Have you ever heard of that, Kate?

Kate said, "Wow. In ancient myths, the Ziggurat was a temple associated with a sort of portal. Portal to other places, other planets. *Nibiru* in particular."

Robert and Bruno gulped.

"What do you mean *'a portal?'*" Bruno asked. "A time machine?"

Kate said, "No. It's a sort of star gate, which can transport people or things to different places, but it's not a time machine. If the stories are true, it can transport people to different planets, even distance galaxies. But no one has ever found a gate yet, or one that we know about. It's all mythology.

"In Iran, there is what is called the *Great Ziggurat of Ur*, a temple; a lot of it still remains. No one is allowed in it. It was believed, in the tradition of the Sumerians, that the gods would often appear there. It's said this is a fabled portal or gate holding the key to unlocking the secrets of the *cosmos*, and sorts of flying machines were seen. The gods appeared to fly into the heavens and fight with other gods. There are smaller ones in Egypt, near some of the temples, but I don't know of any connection to the Sphinx. The Sphinx is so much older than the pyramids, so there could be a connection! We know there is a lot more buried beneath the pyramids and the Sphinx, a lot more to discover.

"The Ziggurat is a type of temple structure, built in ancient times. Most of them are in Ancient Mesopotamia. The Sumerians believed that gods lived in the temples and used a gate system to move around galaxies. The ancient Sumerian writings refer to the gods, the star people. The gods were using people on earth as slaves to mine gold and copper, to repair their own planet's atmosphere."

Katie continued, "Legend has it, when Baghdad was stormed, in the Gulf War, the soldiers headed straight to the ancient libraries and museums and removed a lot of specific items. And the Ziggurat was secured by the military. To this day, there is an army camped there, with a large group of scientists on permanent rotation. Some archaeologists think all the Ziggurats and pyramids from around the world were somehow linked together. But no one knows how or why. The only Ziggurats in Egypt were small and in semi-ruins."

Robert said, "If we find this gate, we will have to

see what happens, but first we have to get to the Sphinx. We had better get going."

All three climbed into Akil's truck and set off.

Bruno said, "We will need to know more about this gate and the Ziggurat!"

Akil said that he would not take them to the Ziggurat, only the Sphinx. He said it was too dangerous at the Ziggurat and that he was scared. There was too much happening there, too many strange things not for peasants like him, only gods.

Robert said, "We're going to the Sphinx, so don't worry."

STAR PEOPLE

A dust storm above the desert was slowly rolling in, and the wind howled with fury. They had found a way in, next to the ear. A small crack just large enough to scrabble in, a small opening to a broken stairway. Robert, Kate, and Bruno descended into darkness, the sound of grinding sand and the storm above fading the lower they went.

The air was cooler below. The air was charged with electricity that prickled against their skin. It was a long way down, using ropes and some stairs that hadn't fallen away. At the bottom, the tunnel opened into a vast underground chamber:

"The Hall of Knowledge!" Kate shouted.

The walls glowed faintly, alive with colour. The

walls were covered in gold, and carvings shifted and shimmered in the torchlight. Not static murals, but dynamic 3d images replaying scenes from forgotten history, and lines resembling circuitry.

Pictures of the Hanging Gardens of Babylon, the Pharos of Alexandria (the lighthouse), and tall humanoid beings with animal heads. Strange animals that moved as though alive. Star constellations rearranging themselves into star maps that Bruno or Robert had never seen.

The chamber was stacked with hundreds of parchment rolls, cylinder seals, drawings, gold goblets, jewellery, granite boxes, and maps.

"It looks like a living archive—the history of our world and the ancient world," said Bruno.

In the centre stood the machine, similar to the one in Misty cave. It was unmistakable—the same alien geometry, only far smaller and more compact. A ring of tall crystal pylons circled the machine. The pylons glowed, flickering with unstable light, and the familiar hum filled the chamber.

Kate pointed. "That's it. The second engine."

Robert stepped forward, his compass vibrating wildly in his pocket. He pulled it free, the cube-shaped device glowing yellow. It pulsed in rhythm with the crystals, guiding him. He understood,

He laid his hands on the console. The symbiotic relationship responded instantly, surging with light. The hum deepened. Energy coursed through the chamber, rattling the very walls.

And then—the voice.

"Align them…
Align the crystals…"

Robert gasped. The voice inside his head, calm and steady, was speaking directly into his thoughts again.

This time, Robert asked, "Who … what are you?"

"It does not matter who we are—Anunnaki, the Guardians, the engineers, the star people—we are as one. Align the crystals. Balance the flow quickly, or all your other work will be undone."

Robert's hands began to turn the console ring. One by one, the crystals shifted. The first aligned. The hum steadied slightly. The second slid into place, its glow smoothing. The chamber shook violently as he moved the third. Dust fell in sheets from the ceiling, and Kate grabbed Bruno's arm for balance.

The voice said:

"You are the key. Only you can bring order to the system now."

The fourth crystal aligned, and the vibration grew smoother, steadier.

"Why me?" Robert asked, whispering through gritted teeth, sweat pouring down his face.

"Because you are chosen. Because you hear us. You found the pouch. We are connected!"

The last crystal was the hardest. Energy bucked against him, fighting like a living thing. Robert forced it into alignment, his arms burning.

And then—silence.

The hum shifted into a steady, soothing resonance. The crystals glowed with a calm blue glow.

The voice spoke once more, softer now.

"It is done. The Earth can rest. The system is stabilised. The moon has stopped moving away from the planet; it will reset the weather systems around the globe in hours."

Robert staggered back, gasping. Above them, the storm eased. The desert air began to settle; the world itself was starting to heal, taking a breath of relief.

Kate touched his shoulder, her voice low. "You did it. We did it."

But Robert shook his head. "Not me. It … guided me. But yes, we did it!"

Bruno gave a nervous laugh. Kate's gaze shifted past him, towards the far end of the hall. There, raised on a pedestal of black quartz, was a granite box. Inside it lay the *Ankh,* shimmering with unearthly light, as though forged from living metal. It floated slightly above its base, vibrating in harmony with the newly stabilised crystals.

Kate's breath caught. "The key…"

The walls around them flared brighter, constellations shifting into patterns of gates, doorways, pathways between worlds. The carved figures seemed to turn, their eyes glowing faintly as though acknowledging the Ankh's presence.

The voice whispered in Robert's mind again, its tone darker this time.

"This was only the beginning for your planet; there are others that need help. The key will open the way."

Robert staggered, clutching his head. The glow from the *Ankh* grew stronger, filling the chamber with pulsing light.

Kate stepped closer, her hand hovering just above the box. "Robert ... this is what we were meant to find."

"No," Robert said firmly, "We came here to stop the storms, balance the weather before the governments and military could totally control it. That was our task. The key ... it's for something else. Something out of this world!"

Kate said, "How do we get the key out?"

The constellations above aligned into the shape of a door, glowing faintly. A stargate. But the stargate could not be unlocked without the *Ankh*.

Kate said, "We need to translate the symbols in the correct order to open the box."

The *Ankh* pulsed again, and more star constellations appeared on the walls.

The voice whispered one last time:

"There are more civilisations and planets to help! Hurry."

WHEN THE SKIES OPENED

The three lay on the floor of the hall, exhausted, but excited about what they had done. There was now a true bond between all of them.

Bruno turned on his battered radio set, busily trying to find a station. He twisted the tuning dial, and they heard a crackle and then a broken voice:

"This is Cairo state radio, breaking developments over the Gisa plateau.
Dark clouds are gathering; rain is coming. Rainfall for the first time in nearly 6 months. Heavy rainfall is confirmed over the Sinai Peninsula."

More Static crackled, and then another voice broke in, choked with emotion and excitement:

"France reports rain in Paris.
The River Seine is flowing again, and crowds are gathering on bridges, cheering at the first drops."

Another feed cut in on Bruno's ham radio:

"Rome confirms the world famous Trevi Fountain is flowing once more. Hundreds of people are throwing coins into the water, weeping with joy."

Then came:

*"This is Brazil Radio Free.
The ice and snow are melting. The roads are open again.
São Paulo is free of ice and experiencing some flooding."*

The radio cracked again:

*"This is the BBC, the worldwide service.
Around the world, reports are coming in. From Australia to Alaska, weather systems are returning to normal, wildfires that have decimated Canada are receding, the droughts in Europe have been quenched, and ice packs in Mexico are shrinking. People around the world and economies are getting back to a new normal!"*

Bruno turned off the radio. It was done!

The three of them clung together, laughing and crying all at once. For the first time in months, humanity was free of the dust, thirst, and ice that had bound it. The world was alive again.

In London, MP Helen James was bringing the military to task, making the world governments rethink their green agendas. The new world order would have to face the consequences.

The weather machines and the Moon were again stable. The world had been saved this time! In the Sphinx, in the Hall of Records, the *Ankh* still waited in its granite prison, its glow intensifying.

Robert said, "This time, we're off-world! The question we should all ask is, '*Why does man have to try and play God?*'"

ABOUT THE AUTHOR

Shaun, a dreamer, has a gift of seeing and feeling what others often overlook. He thinks anything and everything is possible in his world.

A successful businessman, artist, and father, his creativity stems from years of living in a creative environment, so he sees life slightly differently.

He finds excitement in the unknown, the mythology of the ancient world, the positive energies in other people, and a desire to dream big!

www.ingramcontent.com/pod-product-compliance
Lightning Source LLC
Chambersburg PA
CBHW060253030426
42335CB00014B/1669